Patterns at t

by Susan Markowitz Meredith

We can see patterns at the beach.

We see chairs.

The chairs make a pattern.

The pattern is big, little, big, little.

We see sails.
The sails make
a pattern, too.
The pattern is yellow,
blue, yellow, blue.

We see toys.

They make a pattern.

It is pail, shovel,

pail, shovel.

9

Look at the shells.

They make a pattern.

It is big, little, big, little.

Look at the sandals.
They make a pattern.
It is stripes, flowers,
stripes, flowers.

Look at the ice cream.
It makes a pattern, too.

We see patterns
at the beach!